RESCUE FROM THE BACKCOUNTRY

by Buck Tilton

**Illustrations by
Marc Bohne**

ICS BOOKS, INC.
Merrillville, Indiana

i

THE BASIC ESSENTIALS OF
RESCUE FROM THE BACKCOUNTRY

Copyright 1990 Buck Tilton

10 9 8 7 6 5 4 3 2 1

Printed in U.S.A.

DEDICATION

For Melissa "Bugg" Gray, Robert "Lindy" Lindemann, and Richard "Pokey" Low, who "rescued" the time for me to write.

Published by:
ICS Books, Inc.
One Power Plaza
108 East 89th Avenue
Merrillville, IN 46410

Library of Congress Cataloging-in-Publication Data

Tilton, Buck.
    Rescue from the backcountry : the basic essentials of / by Buck
Tilton.
        p.   cm. -- (Basic essentials series)
    Includes index.
    ISBN 0-934802-69-6 : $4.95
    1. Wilderness survival. 2. Search and rescue operations.
I. Title: Basic essentials of rescue from the backcountry.
GV200.5.T55 1991
613.6'9--dc20   90-26011
                                                              CIP

ii

TABLE OF CONTENTS

INTRODUCTION

Mount Webster is a steep hump sitting just northeast of Mount Washington and sharing the frequent cold, wet, screaming winds that, to some, form the most distinguishing characteristic of the White Mountains. The terrain of northern New Hampshire is damp, chilly, stony, treacherous, and easily accessible to a multitude of people who walk in unprepared. Rescues are common. On this particular evening a party of three hikers was nearing the summit of Webster when one slipped on a wet rock and refused to walk further on the injury to his right leg. As darkness closed in they wisely set up camp.

One of the three hiked out with a flashlight and the story of the accident that related where the injured man was waiting. He walked back in with fresh food for his buddies. Our team organized a carry and started in just before dawn.

That same dawn the injured man awoke saying he felt much better. The party of three geared up, limped over the summit and down the opposite side of Webster from the one we were laboring up. We never saw them.

Sometimes it happens that way.

Sometimes it's like the night on Owe Creek when we went in after dark because the message we received said a man was dying. The location we were given was wrong, and we stumbled up a trail nowhere near the victim. He was rescued, eventually, and went on to enjoy complete recovery from his injuries. But we had nothing to do with it.

Last winter we searched for days in a storm for a lone man who turned up only recently when the melting snow uncovered his body.

Then there was the night we got a detailed description of the accident, the condition of the victim, and the location of the group. The note from the party needing aid said they would be fine until help arrived. We hiked in at a rather leisurely pace to find hot chocolate waiting for us. No problem.

Daniel, the Webster for whom the mountain was named, defines rescue: "to free or save from danger, confinement, evil, etc." sometimes it works out that way and sometimes it doesn't. But whether successful or not, it is time for the responsibility of a happy ending to be shouldered by the people going into the remote location instead of being placed on the heads of a Search and Rescue (SAR) team who might be available and who might be out of town for the weekend. The wilderness demands self-reliance, or at least the wisdom to know when and how to send for help.

Start thinking about rescuing someone from the backcountry now, in the comfort and security of wherever it is you are reading this.

1. SAFETY SYSTEMS
PLANNING TO AVOID THE RESCUE

Having led several trips into the Utah's Escalante Canyon and surrounding high desert country, I felt confident preparing for this one. Only one problem. All the previous journeys had been in winter when the canyonlands are more kindly disposed toward trespassers. Now it was springtime.

The resulting emergency was trivial in medical terms, except to the victim. She was red and raw from a massive reaction to poison oak. I knew she was remarkably sensitive to certain plants from the medical information sheet she had filled out for me. Further investigation on my part would have allowed me to warn her to stay out of the moist side canyons where run-off creates a happy environment for poison oak.

There was nothing specific for the problem in the first aid kit. It ended up an evacuation for treatment. She was able to return to the field and finish the trip with the group. But I felt terrible. And she felt worse!

So once again my safety system, the one designed to avoid rescues, took a small evolutionary leap. It will probably always be evolving because preparing for a backcountry trip is a dynamic undertaking. The system begins with a well-thought-out emergency plan and continues through application of all the wonderful ideas I come up with.

I approach safety as a reporter preparing an interesting story for publication:

Who

Who is in charge? Is it you? Who shares the responsibility? Have they been in this position of leadership before? Have you discussed what types of emergencies might occur and how each will be handled? Will they follow your lead if the situation becomes tense? Are they trained in backcountry medicine and rescue? Are they familiar with the area you're headed for? There are ideal answers to these questions, but the most important point is: One of your best safety preparations is to share the responsibility with someone you know and trust.

Who are the other members of the group? Are they experienced in this type of outdoor adventure? Are they mentally and physically fit for this journey? Have they filled out a medical information sheet that asks about allergies, medications, and any pre-existing problems? Avoid the Oh-By-The-Way Syndrome - eight days out on a sixteen day trip someone says to you: "Oh, by the way, I'm severely allergic to bee stings and I forgot my anaphylactic shock kit."

No pre-trip decision is more important than who goes with you.

What

What are you taking with you? Are you prepared for the worst possible conditions for that area at that time of year? Are your fellow travelers as well-prepared? Clothing is your first line of defense against a harsh environment. The second line is adequate shelter. If the shelter you are carrying fails can you construct one from the available natural elements? Is your food adequate for the number in the party with extra just in case?

What is the availability of potable water? Are you prepared to disinfect natural water sources?

Heading for the roadhead on my way to the Chilkoot Trail in Alaska one January, I dropped off the group's trip plan at the local office of the National Park Service.

"What's the radio for?" asked the ranger.

"To call in case of emergency," I answered smugly.

"Call who?" asked the ranger. "That valley is too deep and isolated to get a radio message out."

I left the radio.

Whoever manages an area (Park Service, Forest Service, Bureau of Land Management) knows the answers to many important questions.

When the possible environmental hazards (such as poison oak) are known, when the potential group medical problems are known, the gear and clothing list can be made and the first aid and emergency kit can be prepared for each trip. Each participant should be briefed on what emergencies could occur, how they can be prevented, and what is available to deal with each potential problem.

What you take with you can be almost as important as who goes with you.

When

Although you don't want to remove all chance of spontaneity from your trip, a general itinerary is a healthy idea. A copy should be carried with you, and a copy left with someone who cares about you. An itinerary should include your way in, when you plan to be where, and your way out. It should include the types of activities you plan to be involved in along the way. If you don't show, someone can come looking for you.

Where

Get out the topographical maps and study them. Plan your route and alternative routes. Mark escape routes, the shortest ways to get out to help if needed. I place a "P" where phones are available, and on the back of the map I write the phone numbers of those people who might render aid.

Compare the gear and clothing list and the fitness and experience level of the group to the terrain, activities, and weather conditions expected. Does your planned trip have the highest possible potential for safety?

2. THE ANATOMY OF A RESCUE

One young woman threw her heavy backpack up toward her right shoulder. The mighty effort was successful, but her unbalanced posture slammed the weight of her body and her pack on to her right knee. Several of us heard the joint pop. As we watched it swell, her comment was: "Oh, by the way, I have a trick knee." The nearest road lay a dozen rugged miles away.

The Event

Despite elaborate precautions, fate will sometimes demand a rescue. When the event occurs your first actions should be to maintain a safe scene (prevent a second victim) and treat your patient. Leadership is best expressed in a relaxed and authoritative manner. The more seriously injured your patient the most valuable your efforts to keep the group calm and busy. When the patient is stabilized, the time has come to sit down and have a chat with yourself and the others in your party with whom you share leadership. Chaos may make it seem as if urgency is required, but twenty or thirty minutes to think things through will not make a difference in the health of your patient if the rescue is going to take hours or days. And those thoughtful minutes could be the most important in determining success or failure.

6

Is This An Evacuation?

Determining to evacuate or not is the first and often toughest decision that has to be made. The condition of the patient is the primary factor involved. Is your patient:

1) Trivially injured and able to continue? What is the comfort level of the injured person? Do the planned activities demand more than the subject is now capable of? Are the rest of the group supportive of the patient?

2) In need of a physician's care, but not immediately. Is the patient able to walk out? If not, maybe you can set up camp, fish a bit more, go for the summit, and evacuate leisurely the next day.

3) Critically injured. You'd better be spending your thought time figuring out the quickest, safest way to get your patient to a doctor.

4) Dead. Even though there is absolutely no reason to rush, everyone feels like something should be done right away. The most important first step for you as the leader is to calm the group down and help them deal personally with the situation. The body is usually best left secured for someone else to come in for later. Most groups deal poorly with carrying their own dead, and it is seldom worth the risk of injury to others in order to remove a body immediately.

5) In any of the above conditions, but not easy to access or remove from the backcountry. If you are untrained, you will want to go for expert help. (See **SPECIAL RESCUES**)

Go For Help Or Do It Yourself?

Whether the patient is a member of your party or someone you find in need, this decision is based on a number of factors that include:

1) How fast does the patient need to see a physician, and which method will get him or her there the soonest? Are you able to care for the patient's needs during a carry or would quietly waiting for aid be better for their health?

2) Are skilled rescue personnel required to extricate the victim from his or her predicament?

3) What sources of aid are available at the roadhead? A helicopter? A horsepacker? A rescue team?

4) What is the terrain like? Is it reasonable to think you can carry an injured patient down that mountainside?

5) How soon will it be dark?

6) What is the weather doing?

7) Do you have enough people available to carry a litter, and are they physically capable of the job? The National Association of Search and Rescue (NASAR) recommends litter teams of six-to-eight people and a minimum of three teams. Other people may be necessary to carry extra gear out.

8) Do you have the materials and skill to build a litter if you decide to self-rescue? (see **LITTERS: AMBULANCES OF THE BACKCOUNTRY**)

Your decision should provide safety, with the greatest good being done for the greatest number.

Going For Help

Most of the time you'll choose to go for assistance. But first do nothing. After you've done nothing for a few minutes your mind will begin to clear and some thoughts will have come that should include:

1) I need to write a complete note to rescuers that tells exactly who the patient is, age, sex, what happened and what forces were involved, the chief complaint of the patient, anything pertinent discovered during a physical examination of the patient such as other possible injuries, several sets of vital signs taken over the last 30 - 40 minutes, what has been done so far, and ...

2) I need to choose the nearest suitable campsite, move there, mark it well on a map, send the map out with those who are going for help, and not change camps while they are gone.

3) I need to send, if possible, three people out for help, and I need to be sure they know the way to go and what to do. I need to know that they are as well-equipped as possible for the walk out. I need to know they have the note to rescuers and the map.

4) I need to make the camp as comfortable as possible for the patient, and as visible as possible for incoming rescuers by building a fire, hanging out bright colored clothing or gear, or creating a large "X" on the ground where a helicopter might land (See **SPECIAL RESCUES**).

5) I need to provide the patient with the best possible medical attention during the wait for rescuers.

6) I need to encourage the well-being of everyone in the group with attention to their body heat, hydration, nutrition, and state of mind.

3. SIMPLE SELF-EVACUATION TECHNIQUES

On the streets victims are rushed to the hospital where expert advice and care are immediately available. In the backcountry victims should be allowed a period of rest which gives the injured body a chance to calm down, stabilize, and deal with the harm that has occurred to it. Rushing also predisposes the victim to a worsening of the injury, and sets up a situation that may create a second victim. These are both unacceptable.

Walk-Outs

You may discover your patient has experienced a relatively trivial mishap, a minor upper body injury or some lower extremity injuries, which will allow the victim to walk out if the weight of his or her gear is shared among the group. Given time to stabilize, the condition of the victim is usually best determined by the victim. They know how they feel compared to how they used to feel, and they will best know how usable the injury is.

In all rescues after safety the highest priority should be given to patient comfort.

One-Rescuer Carries

For disabling yet still relatively minor injuries, such as a sprained ankle or an unusable knee or a fractured but stable lower leg, the party may evacuate the patient by sharing not only the weight of the patient's gear but also the weight of the patient. The simplest techniques involve one rescuer carrying one patient, and require nothing more than physical strength and perhaps a few materials.

Patients can be moved short distances easiest with the age-old piggyback method. Carrier and carried soon tire, and longer distances will ask for a better system of holding the patient in place than strength of arm.

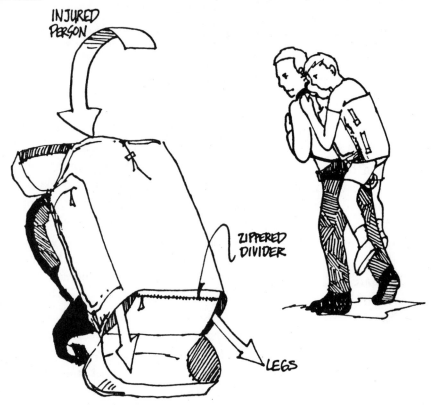

Figure 3-1. The Backpack Carry

The Backpack Carry

Internal frame packs, the kind with a sleeping bag compartment in the bottom, can be slipped on like a pair of harsh shorts allowing the patient to be "worn" out of the backcountry in relative "comfort". This method will insist on much thought if the expensive pack needs to be cut to fit the patient.

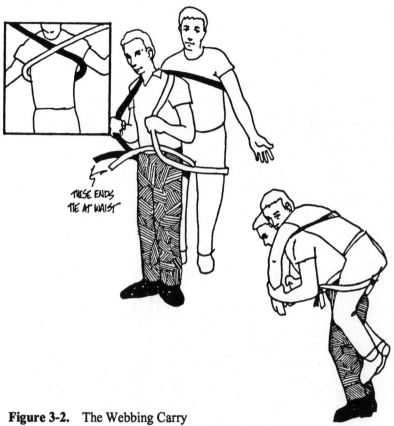

THESE ENDS
TIE AT WAIST

Figure 3-2. The Webbing Carry

The Webbing Carry

With 15 to 20 feet of nylon webbing a carrying system can be improvised. The center of the length of webbing should be placed at the center of the patient's back and brought under his or her arms and crossed over the chest. The webbing then passes over the rescuer's shoulders and back around the rescuer and between the patient's legs. When brought around the patient's legs and tied in front of the rescuer, a seat is formed. When slack develops in the system the webbing can be untied, the patient hitched up, and the webbing tied again. Pressure points in the system, such as the armpits of the patient and the shoulders of the rescuer and the legs of the patient, should be padded to increase the distance the system can be utilized before both participants collapse in pain.

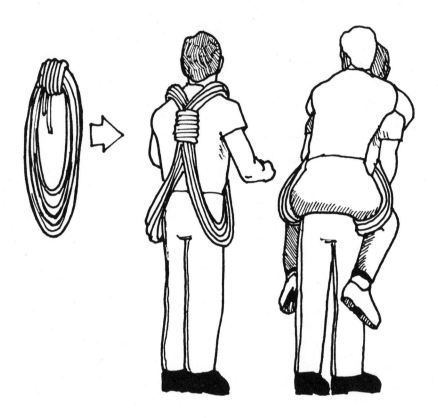

Figure 3-3.   The Split-Coil Carry

The Split-Coil Carry

If a climbing rope is available it can be tied in a mountaineer's coil and split into two approximately equal halves connected at the knot of the coil. When laid on the ground it looks something like a ropey butterfly. Each of the rescuer's arms goes through one wing of the butterfly allowing the coil to be worn like a rucksack. Each of the patient's legs goes through the lower part of the wing allowing the patient to be carried like an awkward backpack. Once again padding at the pressure points adds comfort to the system.

Figure 3-4. Two-Person Carries

Two-Person Carries

When the strength of two people is required to move a heavier patient the weight can be distributed evenly by the rescuers standing on both sides of the patient. Each rescuer reaches for the other rescuer's hand under the patient's arm and behind the back. With the other hand each rescuer reaches under the patient's knees. Using their legs and not their backs the rescuers stand with the patient seated in a temporary chair.

If one rescuer is obviously stronger than the other the weight of the patient can be distributed unevenly by having one rescuer stand at the head of the patient and the other at the feet. The first rescuer reaches beneath the patient's arms and the second beneath the patient's knees. Again the effort to lift is made with the legs not the back. This position is not especially comfortable for the patient over long distances.

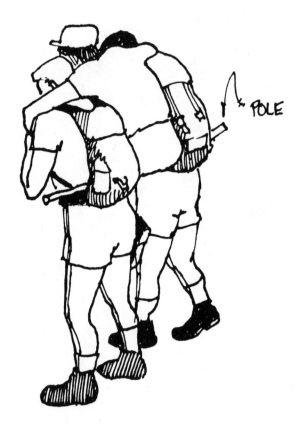

Figure 3-5.   Two-Person Pole Carry

Two-Person Pole Carry

Better is the Pole Carry (or Ice Axe Carry). If two rescuers are wearing rucksacks or backpacks, a pole or long ice axe can be shoved into the strap system or tied to the bottom of each backpack. When the pole is well-padded the patient can sit with relative comfort with his or her arms over the shoulders of the rescuers. The rescuers can carry in relative comfort. This system is awkward if the rescuers are of significantly different height.

None of these carrying techniques are recommended for seriously injured patients. And neither do they represent all the possible ways for one or two rescuers to move a patient. In all rescues you are only limited by your imagination and the comfort level of the patient (and the rescuers).

4. LITTERS: AMBULANCES OF THE BACKCOUNTRY

Some patients, even though seriously in need of care, are generally best treated by camping and attending to their needs as best as you are able while a team goes for help: injuries to the head, the neck and the rest of the spine, multiple fractures that you are unable to adequately stabilize, heart attack, or any other injury that causes the patient a significant increase in pain when you try to move them on a makeshift litter. You may greatly endanger these people trying to move them too far without adequate means.

The most important reason to consider an immediate self-evacuation in an improvised litter is a patient whose condition is steadily deteriorating. These conditions include: increasing difficulty breathing, unconsciousness for an unknown reason, insulin shock or diabetic coma, persistent acute abdominal pain especially with a fever, and life-threatening problems brought on by extremes of cold or heat. You may choose to self-evacuate other patients, whose lives are not at stake, if your rescue promises no further harm to the injured.

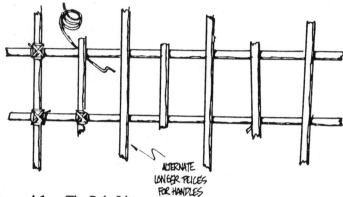

Figure 4-1.   The Pole Litter

The Pole Litter

As old as injuries, the pole litter is made from materials provided by the backcountry. It starts with two poles longer than the patient and preferably of dry dead wood. Green wood flexes horribly making the carry difficult for rescuers and the ride miserable for the patient. Shorter pieces of wood are lashed parallel to the longer poles until a bed is created. Lashing materials include vines, grasses, rope, cord, string, boot laces, or anything else long and flexible enough. For patient comfort the bed needs lots of padding.

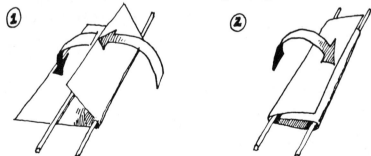

Figure 4-2.   The Blanket Litter or Tarp Litter

The Blanket Litter or Tarp Litter

With the advent of weaving, large pieces of cloth became available to improvise litters. Blankets aren't carried anymore, but tarps are, and a simple wrap around two long poles (as shown) creates a workable litter when the patient's weight binds the tarp to the poles by way of friction. Cross braces lashed to the ends of the poles add stability and ease of carrying.

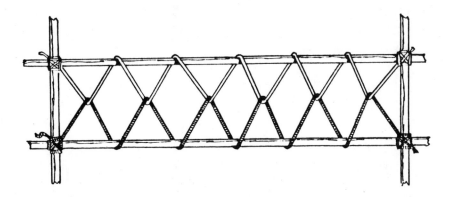

Figure 4-3.   The Pole-and-Rope Litter

The Pole-and-Rope Litter

Something like a hammock can be woven between two poles if you have enough rope on hand. Two long poles are laid on the ground about shoulder width apart. Crossbars are needed at each end to give the litter stability. The crossbars can be sticks, ice axes, tent poles, anything available. With one end of the rope lash one end of one crossbar into place. Wrap the rope 373 around and around one long pole pulling the loop formed by each wrap out to an imaginary center line between the two poles. You can cut the rope and lash the second crossbar into place on one end, or lash and continue with the rope whole. Lash the other end of the second crossbar into place and begin to weave the rope around the second long pole and back through the loops you've made from the other side of the litter. The number of loops is completely relative to the length of your rope, but the more loops the more solid the patient's bed. Take all the slack out of the system that you can and tie off the remaining end of the first crossbar. Pad the bed very well.

The Packframe Litter

Many people still prefer to use an external frame pack and three or four of those frames can be lashed together to make a very substantial litter. Since the frames are molded to fit the shape of the wearer's back they should be placed together in a way that best matches the shape of the person who will be carried. Take time to ensure the lashings are secure. Improvised litters can rattle apart surprisingly fast during a carry. And once again patient comfort demands much padding.

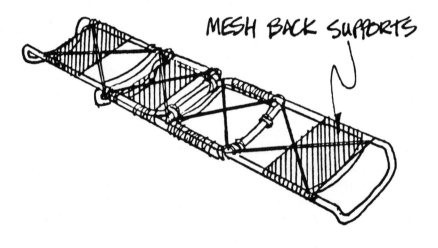

MESH BACK SUPPORTS

Figure 4-4.   The Packframe Litter

The abundance of straps on internal frame packs often allow them to be connected to two long poles to form a litter. Without a lot of straps you might consider cutting holes in the packs at functional places in order to slip the poles through.

The Rope Litter

There are many ways to weave a rope into a litter. This one works best if you start with an approximately 150-foot length, find the center, and stack the two halves ready for use. From the center make

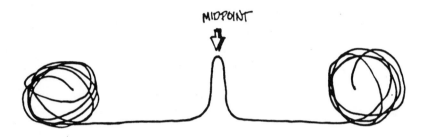

Figure 4-5.   The Rope Litter, Step 1

14 - 18 bends in the rope, 7 - 9 of them on either side of the center. The bends should be about as wide as the patient. If you have a full-length Ensolite pad you can lay it on the ground first as a pattern for the litter. When the bends are finished, bring the rope down the side of the bends tying a clove hitch in the rope at each bend and pulling a loop of the bend through the hitch until all the bends are looped through clove hitches. Continue around the litter with each end of the

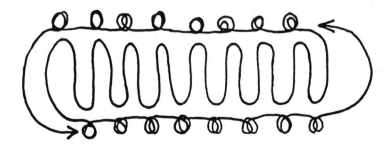

Figure 4-5.   The Rope Litter, Step 2

rope, pulling it through each loop you have created until all the rope has been used to make the sides of the litter. Tighten the knots and tie off the ends of the rope. Pad the bed.

If two long poles are available they can be pushed through the loops before tightening the knots to add rigidity to the litter. If cross braces are added to the poles a sort of hammock is formed for the patient.

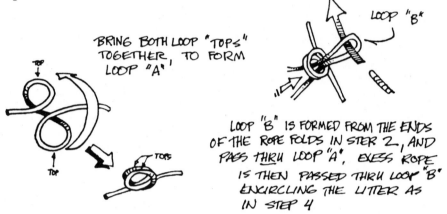

Figure 4-5.   The Rope Litter, Step 3

Any time an improvised litter is used a healthy person should be placed in the litter first to test it for strength and comfort.

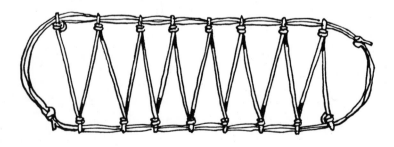

Figure 4-5.   The Rope Litter, Step 4

The Commercial Litter

Several manufacturers offer litters made from metals or plastics or fiberglass or combinations. They all have the same general shape. Many of them can be attached to large wheels and pushed like a cart, or converted into sleds and pulled over snow, or attached to hauling systems and lifted by a helicopter or dragged up a vertical slope. Most commercial litters are much easier to handle and carry than improvised litters. And they all have one big advantage over improvised litters: they provide better patient care.

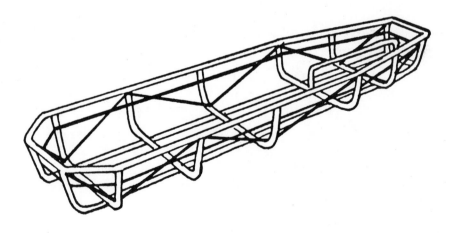

Figure 4-6. The Commercial Litter

5. PACKAGING AND CARRYING

Riding in a litter is at best an uncomfortable experience for the patient. All the attention you can pay to the details of packaging a patient for a carry is time well spent.

Most patients are packaged on their backs because better care can be given to their injuries, and they fit better. In relation to comfort it is probably impossible to put too much padding on the bed of the litter. Soft pads behind the knees and the small of the back ease the discomfort of a long ride. If you are not worried about the possibility of a neck injury provide a pillow for the head. Even if a neck injury is suspected some slim pad beneath the head is a must.

Some patients will require packaging on their side. This person is usually unconscious and you are concerned about maintaining an adequate airway which the stable side position can offer.

Whether on their back or side all patients need to be secured in the litter. This means strapping them in. Even the smoothest trail is a rough ride for the patient and the last thing you want is for them to slop around or out of the litter. Staying in the litter is undoubtedly high on the priority list of the patient as well. The straps should be placed on the bony structures of the patient: lower leg, upper leg, pelvis, chest. Do not place straps across the abdomen or neck. Straps across the chest should not impede breathing. The arms should be left

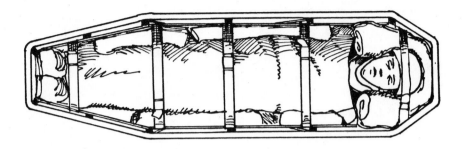

Figure 5-1.   Securing the Victim

free if the patient is conscious. Freeing the arms increases the patient's feeling of security, and decreases the claustrophobia inherent in a litter. The tightness of the straps will depend partially on the amount of pain they cause the patient, but you should be able ideally to roll the litter on its side without the patient falling out (or the litter falling apart). Pad well beneath the straps where they cross the patient.

If the walk out involves some steep terrain the patient will appreciate some sort of step to brace one or both of his or her healthy feet against. Tie a piece of nylon webbing or rope to a side of the litter and tie a loop in the other end into which one of the patient's feet can be placed. When the litter turns downhill the patient can "stand" in the litter. When patients are unable to support themselves with their feet in steps, tie a loop of some material around their thighs and to the side of the litter to form a seat that will support them if they start to slide downhill. Pad where the seat presses against their flesh.

Standard of care demands that a patient suspected of cervical spine injury be provided with a cervical collar. One can be improvised from clothing or by cutting the end off an Ensolite pad and wrapping it carefully around the neck. The collar should prevent as much head movement as possible without strangling the patient. With the collar in place the patient's head must also be secured to the litter. This is best accomplished if a lot of padding is placed around the head and a strap is secured over the forehead compressing the padding snugly against the sides of the patient's head. Conscious patients are often unhappy about this treatment, but it has saved many lives by preventing dangerous motion of an unstable neck.

In wet conditions place a tarp or sheet of plastic in the litter first. When the patient is in the litter fold the covering into an adequate waterproof shield and secure the patient to the litter. Small slits can be made in the rain covering if steps are required for the patient to stand in.

If it's cold the patient does not have the advantage you enjoy of being able to exercise to increase personal warmth. Great attention should be paid to keeping a patient warm. Surround the patient with all the insulation possible but without taking away clothing from other members of the party. Cold can easily create a second victim. If the patient is in a sleeping bag be sure it is zipped up. All the places you feel cold creeping in when you're sleeping out, the patient will feel even more due to their injury. Be sure the head is warmly covered. If you have heat packs, or have made hot water bottles, do not place them directly against the patient's skin, but do place them at critical spots: feet, hands, neck, chest, perhaps the groin. Once again use the tarp or sheet of plastic as a final barrier against the cold. It will work as wind- and snow-proofing as well as a radiant layer reflecting the patient's body heat back where it is needed. Carefully tuck the

Figure 5-2. Carrying the Victim

covering in around the patient. Cold sneaks in through the smallest opening. The straps go over the cocoon you have created to secure the patient to the litter.

Litter bearing provides a wonderful opportunity to create a second victim. When it is finally time to move the loaded litter remind the bearers to lift with their legs and not their backs. Space the bearers out evenly, three to each side of the litter. If possible match a team of bearers up according to height. Encourage constant quiet communication between the bearers. Remember the people carrying the back end of the litter have difficulty seeing where their feet are being placed. A scout should precede the litter picking the optimal pathway and calling back "road conditions" to the bearers. Any remaining members of the party are best kept behind the litter and out of the way.

Fatigue comes quickly to the litter bearer. Fifteen minutes is about all a normal bearer can safely bear without a break or a change of bearers. Overtiring a bearer is dangerous for everyone. To conserve the strength and good humor of the party keep the litter moving forward when it's in the air and set it down for discussions or changes of the guard.

Ease of carrying can be increased if 8 - 10 feet of rope or webbing is tied to the litter for each bearer. This strap is thrown over the shoulder of the bearer and held in the hand to the outside of the litter. With this strap the bearers can shift the weight to their shoulders and even completely free the hand holding the litter from time to time.

When the sides of the litter consist of two long poles a backpack can be tied to the front and back of the litter and the weight borne by two bearers. This system creates much strain on the bearers, but it can be used when bearers are in short supply. Better yet is to tie two backpacks to both sides of the front end of the litter and two to the back end. Now the weight can be carried by four bearers with less difficulty, if the terrain is open enough.

Drops in the terrain, fields of boulders, and narrow waterways sometimes create obstacles to progress. For the safety of everyone do not try to carry over these obstacles. Instead bring the remainder of the party to the front of the litter and form a chain to pass the litter down or across the obstacle. If the pass is long, bearers can peel off the back and move to the front of the litter to continue the caterpillar-like motion. Ask the bearers to make sure they have secure footing before they accept the litter's front end and to make sure they have passed the litter to someone securely before they let go of the back end.

On sustained steep terrain safety is best maintained if a rope is

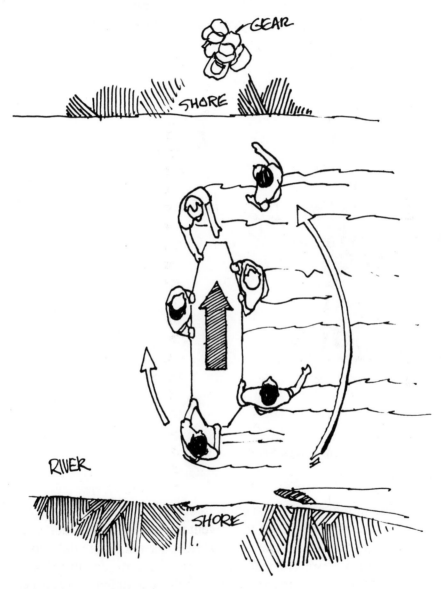

Figure 5-3. Traversing waterways with litter victim

tied to the back of the litter and a running belay is played out from around a rock or tree. In a running belay the rope is simply wrapped once around the anchor (the rock or tree) and the free end let go slowly by a belayer. The belay allows the bearers to lean down the terrain using the litter for support while holding it up at the same time.

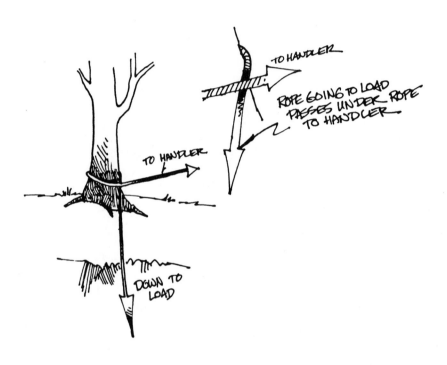

Figure 5-4.   Using belays for descending

If the circumference of the anchor is substantial the friction on the rope will allow the belayer to prevent movement of the litter with little effort should he or she need to. Steep terrain is here defined as one in which a person can walk up or down without using their hands for support. Terrain steeper than this should be avoided. If it can't be avoided, technical rescue systems are recommended (See **SPECIAL RESCUES**).

When you move from one place to another your feet lead the

way. You have become accustomed to following your feet, and so has your patient. Patients in litters almost universally prefer to be carried feet first, even if the terrain slopes uphill. Going downhill adds to discomfort and motion sickness.

If a litter-bearing team all stands about the same height the litter obviously rides more level. But less obvious is how much more smooth the ride if everyone carrying the litter attempts to walk in step, when the terrain allows it.

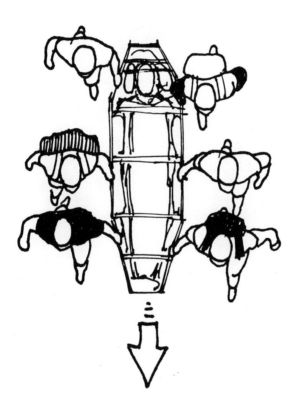

Figure 5-5. Carry Victims Feet First

If night falls, stumbling increases, progress slows, and safety is compromised. Camping is the best option unless the patient's condition and the availability of battery-powered lights ask for a continuation of the rescue. Headlamps are more practical than flashlights. If they are worn around the rescuer's necks instead of their heads all the illumination is pointed toward the ground and the bearers will get fewer blinding flashes of light in their eyes.

As a bearer your back will ache, your shoulder will scream, your arm will stretch painfully out of shape, your hand will cramp, your patience will break and drain into the soil of the backcountry along with your sweat, and you'll agree that extended carries are a pursuit pursued only by confirmed masochists.

6. THE CARE AND FEEDING OF LITTERED PATIENTS

It is very lonely inside a litter. There is pain and discomfort, and sometimes darkness, and the inability to move around to find just the right spot. The muffled sounds that come from the people carrying you are often unintelligible. You feel terrible to cause this much trouble, but there's nothing you can do about it.

As a litter bearer assume the responsibility for the mental as well as the physical comfort of your patient. Talk to the patient, especially when the litter is not being carried. Lean down where the patient's face is and speak quietly and confidently. If someone else is talking to the patient don't butt in. It's difficult to carry on two conversations under normal circumstances, impossible for an injured person. Ask about the injury: What can be done to make it more comfortable? Is it better, worse, the same? Ask about the amount of warmth inside the litter: Too little? Too much? Answer the patient's questions honestly and to the best of your ability.

If the patient is unconscious talk to them anyway. Most often they are not nearly as far away as they seem. Reach inside the covering of the litter to check for warmth. Listen to the breathing and check the quality of the pulse gently at the carotid artery in the neck. Unless it is blistering cold you may choose to periodically uncover an unconscious patient to make sure any splints you've put on are not cutting off circulation.

In terms of bodily needs the basic rule of the rescue is: If you gotta, they gotta! When you drink, offer them a drink, if the injury allows. When you eat, try to get them to eat, again if the injury allows. If you feel a chill coming on, even more so the possibility that they are. If you are too hot, check the patient's comfort.

When nature calls you to the bushes for a break it's time to check in on your patient's need to urinate. They are often too embarrassed or too unwilling to cause a commotion to ask. You might have to unwrap your neat litter package to accommodate the patient, but that's part of the game. The job can usually be accomplished for a man by turning the litter on its side if the day is warm and he isn't too deep in coverings. Or let him stay on his back and use a water bottle as a urinal. The problem is less easily solved for a woman, but she can sometimes catch most of her urine in a wide-mouth water bottle. If something absorbent is placed between her legs first any spill can immediately be caught. In the worst scenario you have to take your patient out of the litter and let them do the job and repack them.

It is a sad truth that too many rescuers are too much in a hurry to get out of the woods at the expense of the patient. If they have a need take the time to see if it can be met.

7. SPECIAL RESCUES

The Avalanche

Most of the avalanches that kill someone occur in the backcountry where there is no organized means of attempting a rescue. The avalanche is usually started by the party that is buried. If someone is to survive burial in an avalanche the immediate actions of the non-buried members of the party are especially critical. Statistics indicate only one-half of those buried will live if they are still under the snow one-half hour after the avalanche. There is seldom time to go for help.

Avalanches slide into two general categories: loose and tight. A loose snow avalanche occurs when the grains of snow are loosely bonded together. The snow is usually freshly fallen, or sometimes extremely wet. The slide starts at one point and fans out as it descends the slope. Loose avalanches seldom cause serious trouble to backcountry users unless they catch them in a narrow ravine where the snow can build up volume and force. Or if the loose avalanche kicks free a heavier body of snow on the way down. Safety requires that any avalanche zone be avoided once recognized.

Tightly held together snow, contrarily, breaks free of the slope all at once as a slab. Deep accumulations of damp snow deposited with the aid of high winds can sometimes develop cohesion and slide as a soft slab during or just after a storm. Most of the killer avalanches slide a few days after a heavy storm when the weather has cleared, the skiers and climbers are out, and the snow has compacted very tightly into a layer that slides as a large hard slab. Approximately one-third of those who die in avalanches are killed by the traumatic impact of the fast-moving hard slab. The other two-thirds die of suffocation waiting for rescue.

If caught you can increase your chance of surviving an avalanche by swimming frantically in an attempt to stay on top of the snow. The records show that shedding your skis and heavy pack will improve your ability to swim. If you are pulled under just before the snow stops moving, shove a hand or arm up across your face to create an airspace. This pocket of air will probably figure immensely in your survival. As a last effort, after creating the airspace and before the snow quickly becomes cement around you, try to jab one of your hands up through the snow. Finally, your chance of living to tell the tale is increased once again if everyone in your group is carrying an electronic transceiver set on transmit. If you are buried, everyone else turns theirs to receive and begins the rescue.

Avalanches start slow, even the huge, destructive ones. There is usually time to run or ski vigorously at a right angle to the slide area and escape before the speed of the avalanche ruins your day -- and maybe your life.

Perhaps you are among the fortunate ones who missed being caught in the avalanche. Make every safe effort possible to keep the victim in sight until the moment he or she disappears under the snow. Begin looking for the victim at that spot, searching first for any sign on the surface: a hand, clothes, ski poles or ski tips sticking out of the snow. If you find clues but not the victim, leave the clues where they are. Their position may be important later. Immediately search uphill from the clues. Their lighter weight probably kept them on the surface after the victim was buried. Form a line with the remainder of the party and probe into the snow with whatever you have to probe with: skis, ski poles, snow shovels, ice axes. Snow flows like water and your search should follow the snow uphill in reverse of the line it slides down. Concentrate your efforts on the outside of curves and uphill of obstacles that may have trapped the victim.

If the horror of the first half-hour passes without success, it is time to go for help. Divide your party into a small group who is adequately prepared to seek aid and the rest who can keep searching.

The Cave

All caves have one great common denominator: darkness. Winds come and go, but the temperature in any given cave tends to stay very constant and varies with each cave depending on the average surface temperature of the geographic area of the cave. The underground passages were cut by the patient flow of water and most North American caves are still very wet. The moisture produces high humidity and the constant threat of hypothermia for the caver.

Caves are rarely simple tunnels. They are most often maze-like, up and down branch upon branch of constantly changing size. The challenge is not in finding your way in, but finding your way out. The most common emergency is a search for someone who is lost. Those involved in the search should be well-trained and well-equipped. If you are untrained in search techniques of cave rescue, you should mark well the last known spot of the lost and go for help.

Once you find yourself underground with an injured victim your problems quickly multiply. As always patient comfort and warmth are critically important but made more complex by the dampness, darkness, and lack of room. Litters often become no more than the patient firmly wrapped in a waterproof covering because there is only space to drag him or her along the low passageways. Tag lines must sometimes be attached to the victim so they can be pulled through extremely small tunnels after the rescuers have crawled through first. In narrow passages that do not let bearers walk beside the litter, the turtle carry is sometimes utilized. The turtle crawls on hands and knees with the litter digging into his or her back. Bearers fore and aft balance the litter and help carry when the cave permits.

If you have prepared wisely for your journey beneath the light of day you are ready to leave your patient as comfortable as possible and go for help.

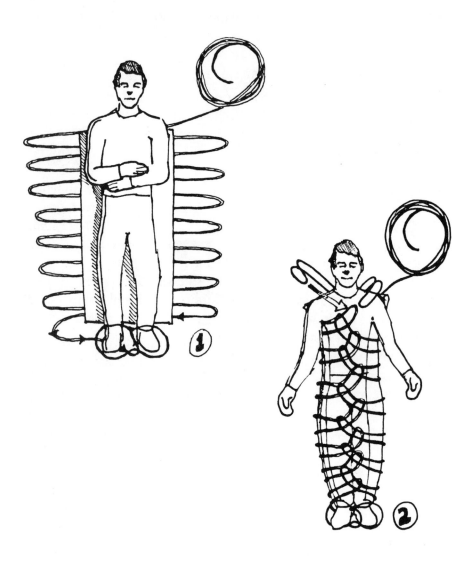

Figure 7-1.   The Mummy Litter

The Mummy Litter

When passages are narrow and a rope and sleeping pad are available, cavers may choose to construct a litter suited for skinny places: the mummy litter. Starting from one end of the rope lay out large bends on the ground. Tie a loop at the end of the rope where you started. Place the sleeping pad on the bends and place the patient on the pad. (Note: the drawing shows a three-quarter length pad often preferred by cavers, but a full-length pad, enclosing the head, provides greater patient care.) Loop the patient's feet, as shown, and bring the bends over the patient consecutively chain-stitching the patient into a cocoon. Include the arms if the patient is unable to control them. The remainder of the rope can be used as a tag line for dragging, or the mummy can be carried.

The Helicopter

Nothing has done more to change the face of rescue than helicopters. They land in remote areas on spots on the earth inaccessible to aircraft only a few years ago. If the spot isn't flat enough they've been known to land on one skid while a victim is quickly loaded. When there is no spot to land they have hovered while rescuers rappelled down and the victim hoisted out.

Helicopters go where the pilot wants because of the rapid spinning of two blades. The large overhead blade(s) creates lift by forcing air down. The pilot can vary the angle at which the blade attacks the air and the speed at which it rotates in order to vary the amount of lift. The entire rotor can be tilted forward, backward, or sideways to determine the direction of travel. Without the second blade spinning in an opposite direction the helicopter would turn circles helplessly in the air. Very large helicopters have two large blades spinning in opposite directions, one fore and one aft, but most of the helicopters used in the backcountry maintain stability with a small tail rotor.

When they are close to the ground the spinning blades build a cushion of air that helps support the helicopter. But this cushion of air varies in its ability to work depending on its density. Rising air temperatures and increasing altitude reduce air density. So trying to land on a mountaintop on a hot day is dangerous for a helicopter. And the weight of one person may eliminate lift-off.

Air density is also altered by the nearness of a mountainside. The downward shove of air by the blades can recirculate off the mountainside and reduce lift.

One of the greatest fears of mountain flying is a sudden downdraft of air than can slam a helicopter toward the ground. Downdrafts are not only dangerous but also unpredictable.

Add to air density and downdrafts the possibility of darkness and fog and wind and you will understand that even if a helicopter is available it may not be able to come to your rescue.

When you're in need of a rescue the approaching thump-thump-thump of rapidly rotating blades is a joyous sound. To give the helicopter rescue the greatest chance of success find a suitable landing zone (LZ). The ideal LZ should not require a completely vertical landing or take off, both of which reduce the pilot's control. The ground should slope away on all sides allowing the helicopter to immediately drop into forward flight when it's time to take off. Landings and take offs work best when the aircraft is pointed into the wind which gives the machine the greatest lift. The area should be as clear as possible of obstructions such as trees and boulders. And clear of debris (pine needles, dust, leaves) which can be blown up by the wash of air and produce mechanical failure. Light snow can be especially dangerous if it fluffs up dramatically to blind the pilot. Wet snow sticks to the ground but also sticks to the runner of the helicopter and adds dangerous weight. Tall grass can be a hazard in an LZ since it disturbs the helicopter's cushion of supporting air and hides obstacles such as rocks and tree stumps.

To prepare an LZ clear out as much of the debris as possible including your equipment and all the people except the one who is going to be signaling the pilot. Mark the landing zone with weighted bright clothing or gear during the day or with bright lights at night. In case of a night rescue turn off the bright lights before the helicopter starts to land. They can blind the pilot. Use instead a low-intensity light to mark the perimeter of the LZ, such as a chemical light stick. Or at least turn the light away from the helicopter's direction. Indicate the wind's direction by building a very small smoky fire, hanging brightly colored streamers, throwing up handfuls of light debris, or signaling with your arms pointed in the direction of the wind.

The greatest danger to you occurs while you're moving toward or away from the helicopter on the ground. Never approach from the rear and never walk around the rear of a helicopter. The pilot can't see you and the rapidly spinning tail rotor is virtually invisible and soundless. In a sudden shift of the aircraft you can be sliced messily to death. Don't approach from a downhill side where the large overhead blade is closest to the ground.

It is safest to come toward the helicopter from directly in front where the pilot has a clear field of view, but only after the pilot or another of the aircraft's personnel has signaled you to approach. Remove your hat or anything that can be sucked up into the rotors.

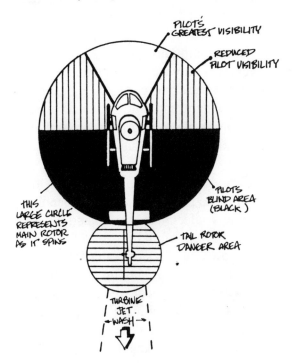

Figure 7-2. Approaching a Rescue Helicopter

Stay low since some helicopter's blades sink closer to the ground as their speed diminishes. Make sure nothing is sticking up above your pack such as an ice axe or ski pole. In most cases someone from the helicopter will come out to you to remind you of the important safety measures.

One-skid landings or hovering while a rescue is attempted are solely at the discretion of the pilot. They are a high risk at best, and finding a landing zone and preparing it should always be given priority. The factors may simply add up to the impossibility of air rescue and alternatives need to be considered in all cases.

The Ice

This is not about the ice that grows on sloping ground requiring crampons and ice axes to move across it safely and ice screws and ice bollards and ropes to move patients across it safely. This is about the ice that grows on top of water providing an excellent opportunity for victims and rescuers alike to drown when they fall through.

When an ice-covered lake is the shortest route and the open space beckons resist the temptation and stay along the shoreline. If the frozen pond or river is the only way without turning back you always have two options: turn back or cross in the safest possible manner.

The safest ice is hard and smooth, clear or ranging from a delightful blue to a shiny black color. A white color usually means the ice has thawed and refrozen, or has never frozen very solid, and may be unsafe.

Noisy surfaces, deep snapping sounds or low rumbles are indications that the freeze is thickening and expanding the ice. This means you are generally safe. On the other hand, when the air warms and ice grows silent, beware of melting.

Melt is also shown by rings of water forming around rocks, sticks, shorelines, or anything poking up through the surface. The surface tends to be weak already near where objects puncture the ice.

Be especially careful of standing water on top of the ice. Water is the only substance known to the universe that gets less dense as it freezes. Ice floats, a fact that makes fish very happy. Ice beneath standing water is often porous from the heavier water working its way down through weaknesses in the ice and making them wider and weaker.

It is best to avoid places where water runs in currents beneath the ice, where streams run in and out of lakes, where springs may be bubbling up in summer, where ice has covered a deep river. Moving water takes a long time to freeze, and it can carve out pockets of weakness underneath a solid looking surface.

When crossing ice, keep your feet wide and slide them along easily. The wideness disperses your body weight and sliding lessens the impact the ice must bear. If you are in a group, keep at least a couple of body lengths between each person. Snowshoes and skis disperse weight even more, reducing the chance of breaking through.

Carrying a pole or the limb of a tree, the longer the better, might prove helpful if you do punch through. You can lever yourself off it to climb out. Without such a pole, stretch your arms as far out on the ice as you can reach and kick madly in an attempt to swim out of the hole and on to firmer ice.

If someone needs your assistance, do not walk out on the ice that they just fell through. Slide them something that floats, or a pole, or throw them a rope. If you feel you must get closer, crawl across the ice spread-eagled, dispersing your body weight to the maximum. But not until someone has tied a rope on to you. When close enough reach for them with a pole and scoot back gently. Or hold on and let them climb out using you as a part of the system.

Rescues from ice-covered water are extremely dangerous and ideally are not attempted without a boat, a rubber raft, or someone wearing a survival suit to keep them high and dry should they fall in. The chance of a second victim is high.

The Lake

Between 5000 and 9000 people will drown in the next year, and approximately one out of every ten of those dead people will have jumped into the water in an attempt to save someone else. The point is simple: in no element is the rescuer more at risk.

There is no doubt about the urgency of getting to the victim in the aquatic environment, but maintaining safety still must remain the highest priority. Safety starts with all rescuers being adequately prepared for the scene. This means they have the appropriate gear, and they have been trained in how to not only rescue someone else but also themselves. There is no excuse for a second victim.

What immediate hazards does the scene hold? How great are the immediate needs of the victim? What equipment is immediately available for use? Who is immediately available to help?

Try to communicate with the victim. If you determine they may be useful in their own rescue, attempt to reach them with your hand, your leg, your shirt, a stick, a paddle, anything they can grasp while being gently pulled to the edge of the water. Although many panicky struggling victims will not hear your cries to them, they will aggressively grab hold of anything they come in contact with.

If they cannot be reached, throw something to them, preferably something that floats. If the something is tied to a line they can be pulled to shallow water. Throw the line past the victim so they'll have to reach for it as little as possible. The pulling should be slow in order to keep the victim's head above water and to prevent the line from being jerked from their hands. Encourage the victim to roll onto his or her back with the line held across the chest.

Paddling out to a victim in a boat is generally a safe method. If they can't be easily and safely pulled over the side, paddle back to shore while holding on to them.

As an absolute last resort, swim out to the victim. This is extremely risky if you have been trained in swimming rescues, and impossible if you have not been trained. Even if your swimming rescue skills are high your chances are better if you're holding onto something that floats or wearing a personal flotation device (PFD) or a wet suit or both. Fear lends the drowning victim the strength of ten thousand. Do not forget that it is OK to wait until the violently struggling victim stops struggling before attempting a swimming rescue.

Three pre-emergency steps will magnificently improve your chances of safely and effectively executing a water rescue. First, decide beforehand that you will not put yourself or anyone at risk during a rescue. Second, keep simple, functional rescue equipment near at hand at all times while participating in water-based activities. Third, learn and practice simple rescue techniques. Don't expect to know what to do by reading it in a book.

The River

Moving water, whitewater, swiftwater. Whatever you choose to call it, rapidly flowing water poses an even greater threat to rescuers than flatwater. Go into that wet world prepared. Everyone wears a personal flotation device (PFD) that fits and offers adequate flotation. Tied to each PFD is a whistle and stout knife. In each boat is a throw bag with seventy-five feet of high-visibility line that floats. Secured to each boat is a waterproof bag with extra clothing, food, a waterproof flashlight, and a first aid kit.

Rapids are stretches of river where the turbulence increases due to a combination of factors that include narrowing of the channel, increases in the slope, the seasonal volume, and the number and size of obstacles in the river. V-shaped tongues of water point downstream between obstacles and generally indicate to the paddler the safest route through rapids. Standing waves sometimes called haystacks form where the water rushes over submerged obstacles. The greatest dangers of the river are (1) strainers, anything in the water that lets water pass through but catches and traps large objects, and (2) hydraulics, or holes, places where the water drops steeply over a rock or other submerged object and curls up and back on to itself. Both of these dangers trap and drown many a swimmer.

There are many tried and often true methods of rescuing victims from whitewater and those who travel there should be as well-trained in rescue as they are in paddling techniques. Here are a few very basic but very important river rules and rescue methods.

Do not attempt to cross swiftwater if the water rises above your knees or if you are unsure of the bottom. Face upriver to minimize the risk of foot entrapment. Lean on a long stick or pole forming a tripod of support. Leave your boots or shoes on. Loosen your pack straps in case you need to get out of it quick. Cross in places where the run-out is safe. If you are washed off your feet you should be sure you won't wash into rapids, or strainers, or holes, or over waterfalls.

If you are caught in moving water after being washed off your feet or out of a boat, do not try to stand up. Swim diagonally across the current to the nearest shore or eddy. If you are swept into rapids keep your feet up and downstream which allows you to fend off rocks and protect your head. Swim with a backstroke for control.

Should you need to throw a line to someone less lucky than you, throw the line as close to the moving victim as possible. Attempt to put the line past them allowing you some margin for error. Neither victim nor rescuer should be tied to this line in case either must release it immediately. Do not attempt to pull the victim to safety. Being towed through water can submerge the face of the victim. The victim should hold on tightly and roll onto his or her back with feet downstream letting the rush of water pendulum them to shore. If instead of shore they are being swept into danger, such as a strainer, they can be released. You run downriver and try again.

Victims entrapped in whitewater pose complex and dangerous rescue operations. Even the seemingly simple task of getting a line secured across a raging river may be extremely risky or impossible for the untrained. Do not go into that cold water unprepared. The result of all rescues should be the same: the most good for the most people.

The Snow

Rescues in snow demand their own special equipment and techniques. As with the aquatic environment a quick response to an emergency saves lives. Snow is, after all, very very cold water. The importance of going into a cold environment prepared for the worst cannot be overstated. Proper gear and clothing, and knowing how to use them, are an absolute must.

Included in the rescue of any winter patient is maintenance of body heat. It is more important to get the patient off the snow than to cover them against the cold, but both are significant.

In the snow it is much easier to drag a patient than to carry them. The most primitive drag method is to insulate the patient well and wrap them in a sheet of plastic, a tarp, a rainfly, and pull them over the snow. On the opposite end of the scale is a covered sled that encapsulates the patient and slides easily on runners. If the sled can be pulled by a snowmobile so much the better.

Figure 7-3.   Improvised Sled Rescue

A very serviceable sled can be improvised from the victim's skis and poles if you have some nylon cord or string and three sticks about two feet long. Carry in your winter pack about eighty feet of nylon cord and a folding saw or wire saw to cut the sticks to length. You'll also want one or two fifty to seventy-five foot lengths of rope to haul the sled with.

Place the skis about two feet apart. Use one of the sticks as a front crossbar by lashing it to the tips of the skis and making sure the lashings are above the snow so they don't create drag. Use twelve feet of cord for each lashing but leave about three-fourths of it free to use later. Use two eight-foot pieces of cord to lash the second stick to the bindings of the ski and leaving about one-half of each piece as a free tail. Do not lash underneath the ski. Place the ski poles on the sticks about shoulder width apart with the baskets over the tips of the skis. Use the leftover tails of the four lashings to tie the ski pole in place but leave enough of the front lashings to form an X from the front crossbar to the opposite ends of the middle crossbar for lateral stability.

A second set of poles, or sticks cut from the forest, are needed to lay alongside the first set of poles extending the bed of the sled to the tail of the skis. Tie the second set of poles to the first set in two places with four two-foot lengths of cord. Tie the third stick to the rear ski poles as a third crossbar. It will take about eight feet of cord to lash securely. Do not wrap the lashings under the skis. Under the weight of the patient the crossbar will sit on the rear end of the skis.

(Note: If the patient can sit up the second set of poles will not be needed. The third crossbar can be tied to the ends of the poles that are already lashed to the skis without using an extension. Under weight these, too, will rest on the skis of the sled with no cord passing under the skis.)

With the remainder of the cord, about twenty-five feet, weave a support net back and forth between the ski poles.

Run the ends of the hauling rope through the baskets of the front ski poles and back and around the middle crossbar and forward to tie them off to the front crossbar. This way when the sled is pulled forward the front crossbar is pulled backwards instead of off the tips.

Pad the sled well, bundle up the patient, and you're off.

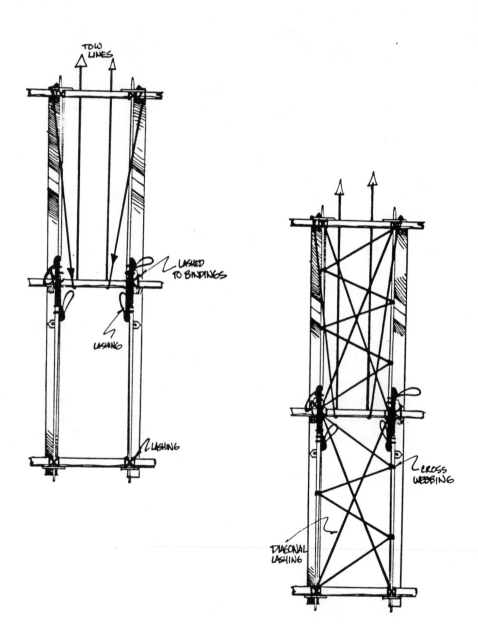

Figure 7-4. Assembling an Improvised Sled

The Vertical Slope

You need your feet and hands to get up and down this environment, and you should know and use ropes and other climbing gear properly to ensure safety. Taking care of yourself and others is a result of training and practice. In no other environment can so much go wrong on a moment-to-moment basis. It could be argued that in no other rescue situation is success more regularly dependent on the equipment and the systems in which that equipment is utilized. Those systems have to adapt quickly to whatever the particular slope offers: bolted anchors, wide cracks, narrow cracks, boulders, trees.

The primary tools of vertical rescue are ropes, and they fall into two broad categories: (1) Dynamic ropes have high stretch and are designed to catch falling climbers with the least impact on their bodies. (2) Static ropes have low stretch and are used when constant tension is put on the system, as in ascending and descending during rescues. Some static ropes are made especially for rescue and have thicker sheaths for greater resistance to damage.

Most vertical rescues involve a variety of specialized hardware that includes (1) carabiners, quick-lock metal connectors used to hook rescue systems together, (2) descenders, metal devices that connect the rope to the rescuer and allow controlled descents, (3) ascenders, metal devices that connect the rescuer to the rope and allow controlled ascents, (4) hardware anchors, an infinite variety of metal devices that fit securely into gaps in the rock to provide safe anchors for the rope systems being used, and (5) pulleys designed to reduce friction on rope systems and increase the efficiency of hauling and lowering.

The personal safety of each rescuer requires the use of seat harnesses which allow the rescuer to "clip in" to the rope system and hang dependent on the rope. And helmets specifically designed to offer protection from falling objects and impacts on the head should the rescuer fall.

Vertical rescuers are skilled in their ability to set secure anchors, tie secure knots, belay (secure a person to a rope in order to prevent damage should he or she fall), rappel (control their own descent of a rope), and ascend a fixed rope.

Before any rescuers put themselves in rope-dependent situations they should be comfortable with the basics of self-rescue. What if they suddenly find themselves hanging upside down in the rope system? What if they catch clothing or hair in the system? Can they improvise a seat harness should the one they're wearing fail? Can they improvise an ascent system with prusiks should the mechanical ascender fail?

Some vertical rescues can be accomplished by one rescuer if he or she has been trained in one-rescuer techniques. These are especially important to master if you climb often with one or two other people.

High angle rescuers know how to secure a litter to a rope system and how to secure a victim in the litter that will be lowered or raised. Depending on the situation they know how to tie the litter in vertically and horizontally. And they are confident about ...

But this does not pretend to be a manual on vertical rescue. This is simply a strong encouragement to get trained and practice, practice those skills. Or go for professional help when a vertical slope separates your patient from safety.

APPENDIX A

Searching

The words "search" and "rescue" are used so often together many people have never learned they describe two separate operations. Rescue involves extracting a known victim from a known environment, and includes the utilization of the quickest and safest method of getting the patient out of danger and into the hands of emergency care providers. Search involves finding someone whose location is unknown.

Another misconception is that rescues are emergencies and searches are long, drawn-out, no-hurry affairs. Conversely, a search is often more of a hurry-up situation than a rescue. Since the subject of a search has an unknown location, finding him or her quickly may be the difference between life and death.

As with rescue, searching has developed into intricate systems, each designed to handle specific lost-victim problems. A search may include hundreds of searchers in a variety of potentially unsafe terrains. But searches most often employ a few people in a relatively small area.

When someone is reported missing, the first step should be to take a deep breath and the time to determine the urgency of the situation. To do this you must gather information from available sources: friends, relatives, locals, authorities. How many people have been missing how long? What is the age, sex, experience level of the missing? What equipment and clothing were they wearing and carrying? Do they have any known pre-existing medical problems? What is the weather doing? What hazards does the terrain pose?

The most common emergency response for a missing person is to dispatch a hasty-team, a limited number of very experienced personnel to take up the search from the last known site along the most likely taken routes. Hasty-teams travel light and fast, carrying all the available information about the victim. Statistically, if the victim if found alive the hasty search discovered the victim roughly 90 percent of the time. As the search evolves into more and more complex operations, the chance of finding the victim alive diminishes.

If a member of your party becomes lost while you're in the backcountry, an organized hasty search is not a bad idea when the rest of the party includes enough experienced personnel to carry it out safely. The hasty-teams should search from the last known site of the victim along the routes most likely travelled by the lost person. The lost, for instance, tend to travel in circles on flat terrain, and generally follow paths of least resistance such as downhill on sloping terrain. If the hasty search has been unsuccessful after six to eight hours, your best bet is to send out for help. As when requesting help with a rescue, send out a written note with as much information as possible relevant to the victim.

APPENDIX B

Signaling

Screams linger briefly in the air and a few tear your throat to ragged soreness. The proverbial needle-in-the-haystack is about as easy to find as a person on the ground waving their arms frantically at a searching aircraft. If you are the lost person or a member of a party awaiting rescue, a few basics on signaling may be important in helping you be located.

The idea is to make yourself heard or seen. Shrill whistles are lightweight, inexpensive, seldom tire the whistler's throat, and the sound carries well. They should be in everyone's pack or pocket. If you're lost, stop moving before you start blowing. Moving targets are harder to hit.

Radios and electronic locators are expensive and seldom carried by lost people. But even with these emergency signaling devices, you should be prepared to augment your visibility. At night small, lightweight strobe lights can be seen for miles if set to flashing from a high point. During sunlight hours, once the rescue party is in sight, any mirror can be used to flash a signal at the rescuers. Stretch out your arm and point at the target. Hold the mirror near your face and cast the sun's reflected light on your outstretched hand while sighting down your arm toward the target. This gives your flashes the best chance of being seen.

To improvise build a fire. These are especially visible at night, but they can be very effective during the day if they are smoky. Once the daylight fire has been generated, wet wood and green vegetation offer a lot of smoke.

During the day strip signals on the ground can be very effective. To make them most effective consider (1) location, (2) size, (3) contrast, and (4) shape. If located on a ridge or in the center of clearing they are much more visible from the air than if they are down in a valley or on a steep hillside. If large, of course, they are more visible. Try to make strip signals a minimum of 18 to 20 feet long. If colors are used that contrast with the natural background, the signal is more visible. If made of straight lines and acute angles they show up better than more natural-appearing curved lines.

There are four strip signals generally understood worldwide:

Y = Yes, or I need assistance.
X = I need medical assistance.
N = No.
→ = I'm headed in this direction.

APPENDIX C

Surviving

Although some of us just seem to be more naturally calm and clear-thinking than others, all of us can exercise and strengthen our survival tendencies. And nowhere is the ability to survive more important than when someone is ill or injured, in need of rescue, and you are the closest thing to an expert anywhere around.

ATTITUDE: The first and most critical stage of survival is described concisely by two words: DON'T PANIC. Panic is the number one threat to survival in an emergency. It destroys the mind's ability to move the body, and along the mindless path lies destruction.

Take a deep, slow breath. Take a couple. Take several hundred. There is a direct connection between breathing and a calm mind. A relaxed mind is the most important survival tool. We do not have the swiftness of the mountain lion, the strength of the bear, the talons of the eagle, or even the fur coat of the field mouse. Most of us couldn't catch a turnip if it wasn't rooted firmly in the ground. But we do have minds and hands, and we can choose to survive.

Recognize the truth of the predicament and choose to meet it with reason. Decide to make the best of a bad situation. Things are almost always better than they first appear.

The second step toward a survival-oriented mind is made with the acquisition of knowledge. This knowledge implies understanding, and understanding doesn't come with hearing and reading. The understanding that works in an emergency is built on practice. One who has constructed a survival shelter or followed the needle of a compass, knows it can be done when the need arises. Now, don't just stand there brooding -- find something useful to do.

SHELTER: The word "survival" seems to turn the human mind to fire and food. Let's rub some sticks together and gather a few wild edibles! The human body may go, theoretically, forever without a fire, and weeks without food. We may even go several days without water. But on a harsh, exposed night in the backcountry can kill us very dead. Adequate shelter will prevent that. To the one needing rescue it is even more important.

Find a place that will give the patient, and the rescuers, as much comfort and safety as possible. Pick a spot as protected as possible, away from heat in hot environments and wind in cold environments, away from weak, overhanging limbs that may break off and rocks that may fall. Choose a place that will stay dry, even if it rains or snows.

A survival shelter should be no larger than necessary. The smaller it is, the less body energy it will require to stay comfortable. Even if there are only low-lying evergreen limbs to snuggle beneath or an overhang to crawl under, make sure the floor is well-insulated and the participants well-covered. Crunchy leaves and prickly pine needles may be all that's available, but they work. The things done for the healthy must be done even more carefully for the sick or injured person who is using much of their energy to heal.

WATER: The body's need for water is almost constant, and slight changes in a human's fluid level can have virtually immediate ill effects, and life-threatening consequences in a very few days. Unfortunately, the waters of the world suffer a common modern problem: the high probability of contamination. Backcountry water may be safe if it's clear, flowing, lined with healthy-looking green plants, free of scum, and filled with smiling fish. Then again, it may be unsafe. The only way to remove doubt is to bring it to a one-roll boil, filter it, or treat it with disinfecting chemicals. Always carrying some means to ensure safe water is a wonderful idea. In the face of an extreme emergency with no means of water disinfection, I would rather drink the safest-looking water than critically dehydrate.

FOOD AND FIRE: These are seldom a serious problem in a rescue/survival situation, but carrying extra high-energy grub and a book of matches, carefully waterproofed, is another excellent idea.

ATTITUDE: The final ingredient in survival-mindedness is an indefinable something called the "will-to-live," an inner urge to survival, the two slices of bread in a survival sandwich.

Don't panic, and carry and healthy attitude.

APPENDIX D

READING TO TAKE YOU BEYOND THE BASICS:

1. Bechdel, Les, & Ray, Slim. RIVER RESCUE. Boston, MA: Appalachian Mountain Club.

2. Chouinard, Yvon. CLIMBING ICE. San Francisco, CA: Sierra Club Books, 1978.

3. La Chapelle, Edward. THE ABC OF AVALANCHE SAFETY. Denver, CO: Highlander Publications, 1961.

4. May, W. G. MOUNTAIN SEARCH AND RESCUE TECHNIQUES. Boulder, CO: Rocky Mountain Rescue Group, Inc., 1973.

5. National Cave Rescue Commission & National Speleological Society. MANUAL OF U.S. CAVE RESCUE TECHNIQUES. Huntsville, AL: National Speleological Society, 1988.

6. Padgett, Allen, & Smith, Bruce. ON ROPE. Huntsville, AL: National Speleological Society, 1987.

7. Peters, Ed. MOUNTAINEERING: FREEDOM OF THE HILLS. Seattle, WA: The Mountaineers, 1982.

8. Robbins, Royal. BASIC ROCKCRAFT. Glendale, CA: La Siesta Press, 1971.

9. Setnicka, Tim. WILDERNESS SEARCH AND RESCUE. Boston, MA: Appalachian Mountain Club, 1980.

10. Tilton, Buck, & Hubbell, Frank. MEDICINE FOR THE BACKCOUNTRY. Merrillville, IN: ICS Books, Inc., 1990.

11. Vines, Tom, & Hudson, Steve. HIGH ANGLE RESCUE TECHNIQUES. Dubuque, IA: Kendall/Hunt Publishing Company, 1989.

INDEX

NOTES

NOTES

Medicine for the Backcountry
by Buck Titlon and Frank Hubbell

$9.95 Paperback • $13.50 Canada • 160 pages
6x9 inches • index • illustrations
Bookland EAN codes ISBN 0-934802-61-0

Learn to provide acute, emergency care of injured back-country travelers. Develop an approach to providing help to victims of trauma, environmental, and medical emergencies. Acquire the knowledge of performing a proper hasty survey of an injured patient. Learn to interpret and record vital signs.

Obtain the skills needed to stabilize a trauma victim and to treat many common backcountry medical problems. Learn how to construct a medical kit for backcountry travel. Instructions are clearly written for easy comprehension. Profusely illustrated.

About the authors :
Buck Tilton and Frank Hubbell are the chief instructors of the famous SOLO Wilderness Emergency Medicine training center in Conway, New Hampshire.